Landscape Architect Registration Examination

LARE Review Section D Practice Problems

Design and Construction Documentation

Second Edition

Erik Sweet, ASLA

Professional Publications, Inc. • Belmont, CA

How to Locate and Report Errata for This Book

At Professional Publications, we do our best to bring you error-free books. But when errors do occur, we want to make sure you can view corrections and report any potential errors you find, so the errors cause as little confusion as possible.

A current list of known errata and other updates for this book is available on the PPI website at **www.ppi2pass.com/errata**. We update the errata page as often as necessary, so check in regularly. You will also find instructions for submitting suspected errata. We are grateful to every reader who takes the time to help us improve the quality of our books by pointing out an error.

LARE REVIEW, SECTION D PRACTICE PROBLEMS: DESIGN AND CONSTRUCTION DOCUMENTATION
Second Edition

Current printing of this edition: 1

Printing History

edition number	printing number	update
1	1	New book.
2	1	New edition.

Printed in the United States of America

PPI
1250 Fifth Avenue, Belmont, CA 94002
(650) 593-9119
www.ppi2pass.com

Sweet, Erik J., 1972–
LARE review. Section D, Practice problems : design and construction documentation /
Erik J. Sweet. -- 2nd ed.
p. cm.
Includes bibliographical references.
ISBN-13: 978-1-59126-087-5
ISBN-10: 1-59126-087-6
1. Landscape architecture--Examinations--Study guides. 2. Landscape construction--Examinations--Study guides. I. Title. II. Title: Landscape architect registration examination. Section D, Practice problems. III. Title: Design and construction documentation.

SB469.6.S94 2007
712.076--dc22

2006051004

Table of Contents

Preface and Acknowledgments

LARE Review, Section D Practice Problems: Design and Construction Documentation is designed to guide your study efforts in preparation for taking the Landscape Architect Registration Examination (LARE)—so you can successfully take the next step in your career. Successful completion of the LARE will allow you continued growth within the field of landscape architecture as well as prepare you for the jump from an entry-level landscape designer to a licensed practicing landscape architect.

Proper preparation for the LARE is as important as the work you do every day in the office and field. Practice problems should be a critical component of your exam preparation strategy. In combination with the other books in this series, *LARE Review, Section D Practice Problems* will help prepare you for the LARE by increasing your solving speed and assessing your readiness. This second edition has been revised and expanded to reflect the exam subjects announced by CLARB in 2006. The problems and fully explained solutions presented in this book are designed to supplement a well-structured LARE study regime. In addition to providing practice, they present successful problem-solving approaches and insights.

I would like to thank PPI for the opportunity to author this book and its second edition. Many thanks to Sarah Hubbard (director of new product development), Heather Kinser (editor of the first edition), Jenny Lindeburg (editor of the second edition), Miriam Hanes (typesetter), Amy Schwertman (designer and typesetter), and Tom Bergstrom (illustrator) for their hard work and efforts in preparing this and other LARE exam preparation books that will add tremendously to the growth of the landscape architecture profession.

I would like to thank Andrew A. Fox, RLA, Assistant Professor of landscape architecture at the University of Georgia, for his insight and expertise, and for helping me create the sustainable design and low-impact development problems in this second edition. Professor Fox is a registered landscape architect whose professional interests include sustainable design and outdoor learning environments. Professor Fox has been involved in the creation of numerous award winning projects, and holds a bachelor of general studies degree from the University of Michigan and a master of landscape architecture degree from Louisiana State University.

I would like to thank Jeff R. Chamlee, ASLA, for his efforts and time as the technical reviewer of the original manuscript and this book's second edition. Mr. Chamlee has made every effort to ensure that the problems provided here meet the exam specifications.

And finally, thanks to my family, especially my three children—Garth, Isobel, and Aidan—who have been patient during the many evening and weekend hours when dad was working to produce a book that will guide many new landscape designers in becoming licensed landscape architects.

Erik J. Sweet, ASLA

Introduction

About the LARE and Section D—Design and Construction Documentation

Successful completion of the Landscape Architect Registration Examination (LARE) is required for licensure as a landscape architect in the 47 U.S. states, one U.S. territory, and two Canadian provinces that license landscape architects. The LARE is administered by the Council of Landscape Architectural Registration Boards (CLARB). Once licensed in any state, a landscape architect can establish a record with CLARB, which can expedite the reciprocity process when applying for licensure in other states.

The LARE consists of five sections. The three multiple-choice sections (Sections A, B, and D) are administered in computer testing centers across the United States and Canada in January, April, July, and October. The two graphic sections (Sections C and E) are administered in regional testing centers in June and December. PPI's website (**www.ppi2pass.com**) gives information on the availability of LARE reference materials covering all these sections.

The Section D exam consists of 120 multiple-choice problems divided into the following subjects. Approximate problem percentages for each subject are given in parentheses.

- design principles (16%)
- resource conservation and management (18%)
- graphic communication (8%)
- construction documentation (20%)
- materials and methods of construction (38%)

CLARB provides detailed specifications for each exam section on its website. Since the test varies each time it is administered, the number of problems in each subject can shift slightly. But reviewing the exam specifications gives an overview of the subject matter covered and can guide the examinee's preparation.

Examinees taking the multiple-choice sections will be required to select the single best choice for each problem, within the allotted time for the section. Most problems will present four options; occasionally, a problem will have up to nine options. Three hours are allowed for the 120 problems in Section D, so each problem should average roughly a 1.5 minute solution time.

What to Bring to the Exam

For the multiple-choice sections, examinees should bring two forms of ID—one of which should be a photo ID—with them to the exam site. It is also a good idea to bring bottled water and an unobtrusive snack.

Examinees are not permitted to bring reference materials, calculators, pencils, pens, or paper to the exam. Calculators, pencils, scratch paper, and a copy of the *LARE Reference Manual* (which provides the code requirements and design criteria to be used on the exam) will be provided at the exam site. Any other materials that would need to be referenced, such as wood sizing charts, will be provided.

Successful Exam Preparation

The first step in preparing for the LARE multiple-choice sections is to confirm the application deadlines and prerequisites for submitting an application. Most states have an application submission deadline three to four months prior to the exam date, so examinees should be sure to meet all experience requirements prior to the application due date (not just the exam date). Although not all states require professional experience to sit for the exam, the confidence and skill set needed to pass come from having a few years of professional experience in a diverse landscape architecture practice.

Every examinee should conduct many study sessions in preparation for the exam. Problem-solving speed and confidence come from practice. It is always in the examinee's best interest to observe good work habits and exercise sound time management during study sessions. Examinees should become accustomed to minor distractions during study sessions; the testing centers often provide their own share of distractions and discomforts.

The *LARE Content Guide* and *LARE Reference Manual*, which can be downloaded from the CLARB website, are critical exam-preparation tools. The *LARE Content Guide* gives exam specifications and other indispensable information on exam content. The *LARE Reference Manual* contains specific code requirements and design criteria that might apply to any problem. Examinees should be familiar with all sections of the manual and be able to quickly reference, if not memorize, the critical information from each section.

An examinee's local chapter of the American Society of Landscape Architects (ASLA) may offer exam preparatory sessions, which present a good opportunity for gaining first-hand insight into the exam and meeting other people who might be interested in forming study groups.

How to Use This Book

This book is organized according to the LARE Section D exam subjects. The percentages of problems presented for each subject approximate those given by the LARE specifications at the time of printing.

To optimize the benefits of this book, readers should first work through all the practice problems, selecting the most-likely answer from among the options presented. When finished, readers can analyze the results by checking the solutions in the back of the book, comparing the given solutions against their own problem-solving approaches and assessing whether their approaches follow the same logic and reach the same conclusions. Based on this analysis, readers can identify personal weak areas and tailor the next phase of study to address those knowledge gaps.

Note that SI units are given in square brackets throughout this book.

Problem-Solving Strategies

The LARE problems change from year to year. However, the nature and format of the problems will be similar across all exams, and getting familiar with the problem types is an effective way to prepare.

Quantitative problems encountered on the LARE multiple-choice sections will require examinees to determine the correct answers by performing calculations. Nonquantitative problems will fall into two general categories: (1) "basic recall" problems, which require examinees to name, identify, or remember the correct term or concept from a list, and (2) "complex" or "situational" problems, which require examinees to apply a principle, concept, or skill; make a judgment; or otherwise address a complex situation.

To solve a basic recall problem, an examinee should start by reading the problem statement only and should then formulate an answer before looking at any of the given options. This approach promotes better recall and helps avoid confusion caused by seeing similar terms presented in the options. Once the correct answer is deduced, the examinee should read the options and select the corresponding one right away.

Responding successfully to a complex or situational problem requires a specific strategy. Often, these problems will be lengthy and will contain information that requires a recall of basic knowledge. An examinee should read the problem statement and then read all the given options. Next, keeping the range of options in mind, the examinee should seek the correct answer while re-reading the problem and then use a narrowing-down process to select the correct response, eliminating choices that do not appear to resolve the presented problem. Careful reading will help quickly narrow-down the choices to reveal or confirm the correct response.

Incorrect answers on the LARE do not count against the total score, so it is in an examinee's best interest to answer every problem. It is possible to flag a problem for later review or to leave one incomplete. Time permitting, the examinee will have an opportunity at the end of the exam to see flagged and incomplete problems and make another attempt at answering them.

A good, time-conserving approach is to work through the entire exam quickly, answering all problems that can be confidently addressed, and flagging the ones that will require more thoughtful analysis. In the remaining time, examinees can go over the unaddressed problems in a second pass, working through the more complex solutions. If any problems are still unanswered near the end of the exam time limit, it is advisable to make a reasonable guess between the two or three most likely options.

Problems

Design Principles

1. In designing an irrigation system, what units are used to measure the velocity of water in a pipe?

(A) gallons per minute (gpm) [liters per minute (L/min)]

(B) feet per second (fps) [meters per second (m/s)]

(C) pounds per square inch (psi) [kilopascals (kPa)]

(D) cubic feet per second (ft^3/sec) [cubic meters per second (m^3/s)]

2. What type of stone masonry, under normal production conditions, would provide a safe, nonslip surface for an exterior residential patio project?

(A) marble

(B) granite

(C) sandstone

(D) cobblestone

3. According to code requirements, guardrails on a wooden deck or platform must be designed with what maximum amount of space between rails?

(A) 2 in [5 cm]

(B) 3 in [7.6 cm]

(C) 4 in [10 cm]

(D) 6 in [15 cm]

4. According to the Occupational Safety and Health Administration's (OSHA's) code requirements, how high must the railing on a deck or platform be above the adjacent deck?

(A) 24 in [61 cm]
(B) 30 in [76 cm]
(C) 36 in [91 cm]
(D) 42 in [107 cm]

5. The main goals of landscape lighting are to

I. facilitate safety
II. improve legibility of outdoor nodes
III. encourage use of a space at night
IV. illuminate to daytime levels

(A) I and II
(B) I and IV
(C) II and III
(D) I, II, and III

6. What is the most critical component in designing a lighting system for a recreational area?

(A) color rendition
(B) uniformity
(C) shielding
(D) footcandle [lux] level

7. The primary goal in lighting a pathway is to

(A) provide inexpensive lighting
(B) provide a safe travel route through the landscape
(C) uplight the surrounding plantings
(D) provide adequate light for parking

8. When planning and designing a site, what three soil property aspects should be considered?

I. physical
II. site engineering
III. horticultural
IV. agricultural
V. textural
VI. cohesion
VII. depth

(A) I, II, and III
(B) I, III, and IV
(C) II, IV, and V
(D) V, VI, and VII

9. When planting in an urban environment, soils will tend to fail due to overcompaction and poor maintenance, which results in

(A) infiltration loss, reduced oxygen exchange, and reduced soil organics
(B) depth loss, reduced oxygen exchange, and compaction
(C) increased runoff, dirty sidewalks, and broken utilities
(D) increased runoff, increased available moisture, and change of color

10. When designing a planting plan, what key criterion will affect the long-term success of the plants chosen?

(A) the maintenance dollars available
(B) the nursery from which the materials were obtained
(C) the colors of flowers and foliage
(D) an herbicide program

11. A good starting point for determining the hardiness of a plant species for use in a particular region of the United States is to

(A) read the tag on the plant
(B) check the USDA plant hardiness zones
(C) review the specifications
(D) use American Nursery and Landscape Association data

12. The design component common to all types of retaining walls is

(A) slope of backfill
(B) drainage
(C) concrete masonry units
(D) cast-in-place footings

13. During the planting of ball-and-burlap materials, the only pruning that takes place should be the removal of

(A) the main leader
(B) dead and damaged branches
(C) one third of all growth
(D) all branches below 6 ft

14. What coating is acceptable to reduce rust on nails used in an exposed wooden structure?

(A) aluminum
(B) zinc
(C) copper
(D) steel

15. To roughly what maximum height is a landscape architect generally permitted to design a retaining wall without consulting an engineer?

(A) 1 ft [0.3 m]
(B) 2 ft [0.6 m]
(C) 3 ft [1 m]
(D) 5 ft [1.5 m]

16. The height of a retaining wall is measured from the

(A) ground level to the top of the wall
(B) top of the footing to the top of the wall
(C) top of the footing to the weep holes
(D) bottom of the footing to the top of the wall

17. In the design of a fast-food restaurant's drive-through lane, a vehicle's headlight glare must be controlled. What type of plant should be placed along the property line to prevent glare from leaving the property?

(A) low flowering shrubs
(B) pyramidal evergreen shrubs
(C) shade trees with an 8 ft [2.8 m] canopy
(D) all of the above

18. At the entrance to a parking lot along a 35 mph [56 kph] road, which of the following areas is restricted to plantings 30 in [76 cm] tall or lower?

(A) sidewalk planters
(B) parking islands
(C) site triangles
(D) parkway planters

19. At the front and side entrances to a bank where there are three daily shifts of employees, which of the following should NOT be planted adjacent to the entry doors?

(A) evergreen shrubs/trees
(B) canopy trees
(C) ground covers
(D) low flowering deciduous shrubs

20. In the design of a high school, a south-facing, central courtyard plaza will be used as a lunch space and as an outdoor stage. Which type of plant would be MOST appropriate in the courtyard?

(A) ground covers
(B) low flowering shrubs
(C) canopy trees
(D) columnar evergreens

21. Which of the following is NOT a valid criterion for using landscape lighting within a design?

(A) Provide a minimum of 1 ft-candle [1 lumen] of light.

(B) Provide ample light to safely walk through the site.

(C) Orient the fixtures so the source of the light is visible.

(D) Select appropriate fixtures for the application.

22. What type of landscape lighting can legally be installed by a licensed landscape contractor?

(A) low-voltage lighting

(B) architectural high-voltage lighting

(C) sports-field lighting

(D) all of the above

Resource Conservation and Management

23. In the design of a project that includes impervious surfaces, such as a concrete plaza or loadbearing asphalt, the soils have been identified as poor and moisture sensitive. What would be used to separate the native soils from the imported materials?

(A) foam insulation

(B) geotextile fabrics

(C) aggregate products

(D) soil cement

24. For plant health, what is the best season in which to plant most materials?

(A) winter

(B) spring

(C) summer

(D) fall

25. What organization has set the standards for acceptable plant health and vigor?

(A) ANSI

(B) ASTM

(C) ANLA

(D) CSI

26. The practice of using plant materials to stabilize banks and control erosion is known as

(A) drought-tolerant planting

(B) scheme planting

(C) bioengineering

(D) urban forestry

27. Which of the following wood finishes would NOT be suitable for exposed exterior conditions?

(A) acrylic latex paint
(B) transparent stain
(C) polyurethane
(D) alkyd paint

28. For an irrigation system in a cold climate, what key factor must be taken into account?

(A) soil classification
(B) growth seasons
(C) frost line
(D) bearing capacity

29. What type of hole is proper for planting a tree?

(A) long and deep
(B) shallow and wide
(C) same size as the container
(D) mounded

30. During the planting of a ball-and-burlap tree with a wire basket,

(A) the wire basket should be removed from the pit
(B) the twine holding the basket and wrapped around the trunk should be removed from the pit
(C) the twine, basket, and burlap should removed from the pit
(D) the twine, basket, and burlap should be folded down into the pit

31. Given the specified depth of backfill in the following detail and a frost depth of 20 in [50 cm], what is the largest mainline diameter that can be used without risk of frost damage?

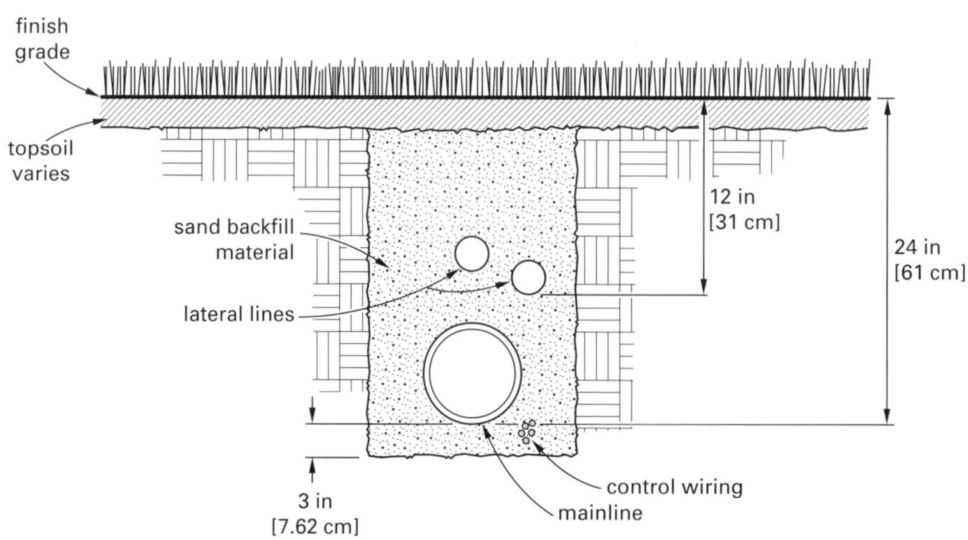

(A) 1/2 in [12 mm]
(B) 3/4 in [1.9 cm]
(C) 3 in [7.5 cm]
(D) 6 in [15 cm]

32. What types of wood treatments are acceptable for a wood post that will make ground contact?

(A) ACX, ADA, CDC
(B) ACA, ADA, CCA
(C) CCA, ACA, oil-based stain
(D) ACA, CCA, wolmanization

33. To remain viable for planting, a stockpile of native topsoil reserved on site for later use should have a maximum depth of

(A) 4 in [10 cm]
(B) 24 in [61 cm]
(C) 4 ft [1.2 m]
(D) 12 ft [3.65 m]

34. The protection of existing trees during construction is best done with

(A) construction tape and signs
(B) construction-barrier fencing at the dripline
(C) an irrigation system, to keep them watered
(D) a mulch ring with a different texture

35. In the plant installation process, what is the most important factor to ensure the long-term health of plants?

(A) size of the planting hole
(B) shipping time of materials
(C) healthy condition of the plants upon arrival at the planting site
(D) proper fertilization during planting

36. Which is NOT a benefit of adding mulch after planting?

(A) It moderates soil temperatures.
(B) It retains soil moisture.
(C) It covers the stem of the plant.
(D) It suppresses weed growth.

37. How long should stakes and guy wires for a tree remain in place?

(A) six months
(B) one growing season
(C) two growing seasons
(D) indefinitely

38. Low Impact Development (LID) is most specifically related to

(A) sustainable management
(B) stormwater management
(C) low maintenance landscaping
(D) all of the above

39. New approaches in sustainable design have grown from various programs and concepts, such as

(A) LEED
(B) LID
(C) BMP
(D) all of the above

40. Associated benefits of a Low Impact Development (LID) approach to stormwater management include

(A) enhanced landscape aesthetics and wildlife habitats
(B) reduced construction costs
(C) increased lot yield
(D) all of the above

41. Changes to a site's existing hydrologic regime typically associated with traditional site development and/or construction include

I. decreased soil porosity and infiltration capacity
II. increased areas of disturbance
III. increased areas of impervious surfaces
IV. increased stormwater runoff velocity, volume, and duration
V. decreased areas of native vegetation

(A) I, II, and III
(B) I, III, and IV
(C) II, IV, and V
(D) III, IV, and V

42. For Low Impact Development (LID) projects, a landscape architect would employ which of the following site design techniques?

I. rain gardens (bioretention systems)
II. porous pavements
III. stormwater discharge structures
IV. detention facilities

(A) I and II

(B) I and III

(C) III and IV

(D) II and IV

43. What do rain gardens (bioretention systems) accomplish?

(A) collect and store stormwater

(B) recharge groundwater

(C) improve runoff water quality

(D) all of the above

44. Which of the following on-site soil recycling techniques contribute to creating a healthy and active growing medium?

I. stockpiling soil in a large pile
II. tilling soil
III. stockpiling and reusing on-site topsoil
IV. covering material with plastic
V. combining soil with on-site trenching spoils
VI. amending soil with certified compost material

(A) I, II, III

(B) I, III, IV

(C) II, III, VI

(D) III, V, VI

45. Which of the following is a main goal of the Low Impact Development (LID) Best Management Practices (BMPs) for stormwater management?

(A) long-term stormwater detention

(B) stormwater collection in a single, large-volume pond

(C) groundwater recharging

(D) sub-surface stormwater conveyance

46. Based on stormwater Best Management Practices (BMPs), the rain garden's soil profile should be made from

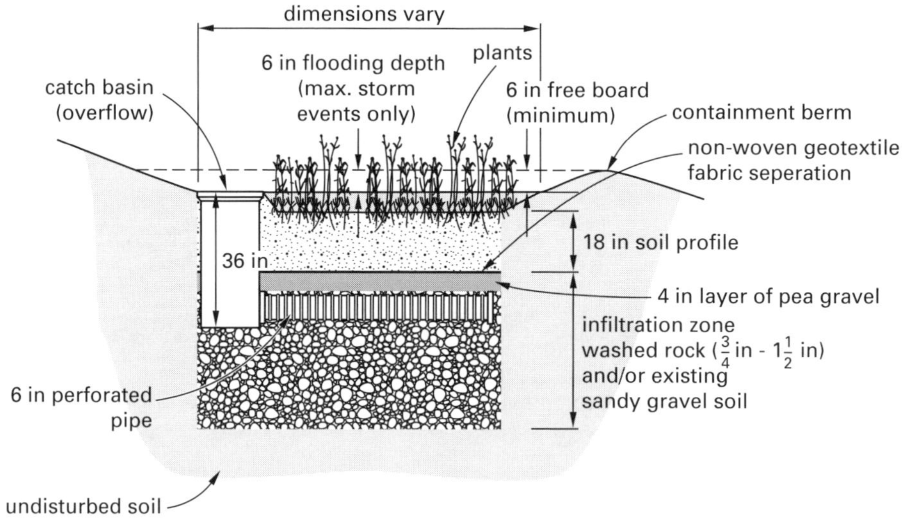

cross section view
(not to scale)

(A) clay soil

(B) concrete sand

(C) compost-amended sandy soil

(D) composted yard waste

47. What types of plants are most appropriate for a rain garden that has no supplemental irrigation? Assume the plants are not drought tolerant.

(A) evergreen shrubs

(B) native vegetation

(C) ornamental flowering shrubs

(D) shade trees

Graphic Communication

48. What percentage of silt, sand, and clay would be in a soil at point A indicated on the following soil classification chart?

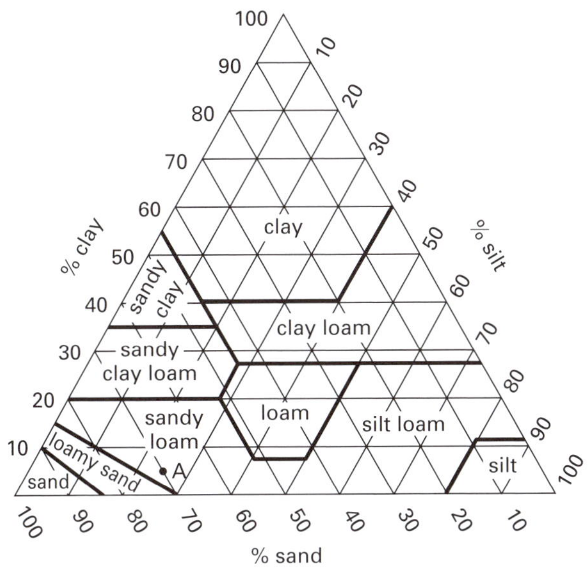

(A) 5% sand, 5% clay, 90% silt

(B) 25% sand, 25% clay, 50% silt

(C) 30% sand, 65% silt, 5% clay

(D) 70% sand, 5% clay, 25% silt

49. What type of fastener is shown?

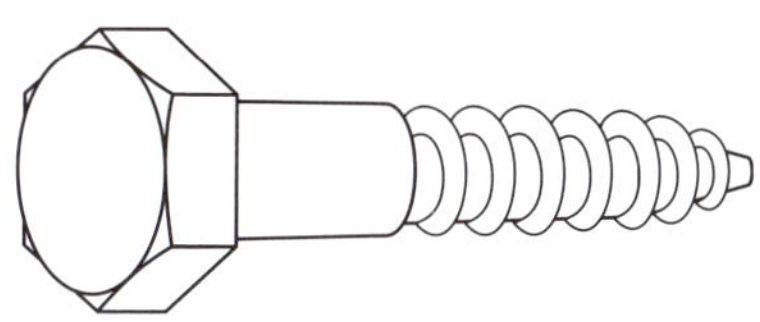

(A) lag bolt

(B) wood screw

(C) machine screw

(D) anchor bolt

50. What type of fastener is shown?

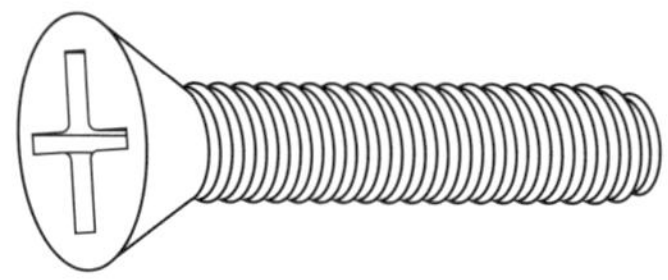

(A) wood screw
(B) lag bolt
(C) toggle bolt
(D) machine screw

51. What type of fastener is shown?

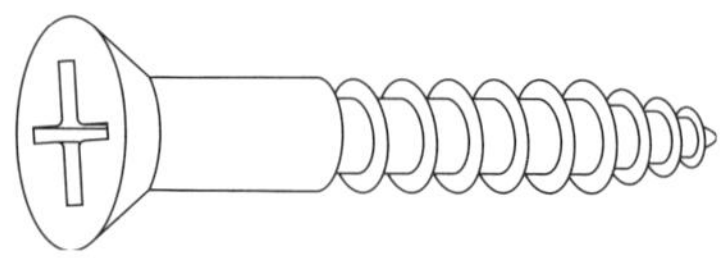

(A) wood screw
(B) anchor bolt
(C) machine screw
(D) lag bolt

52. The fastener shown should be used to attach a

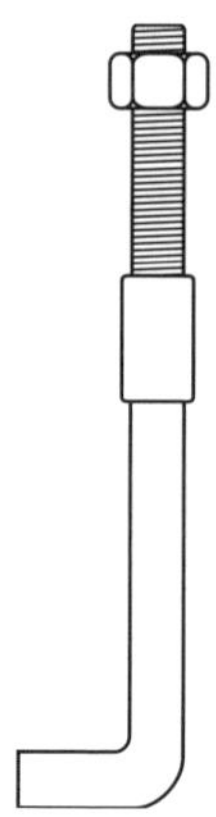

(A) concrete footing to a landscape element
(B) roof truss to a wall
(C) joist to a header
(D) deck header to existing house framing

53. For the following construction detail shown depicting the connection of a joist or a beam to a post, which statement is true?

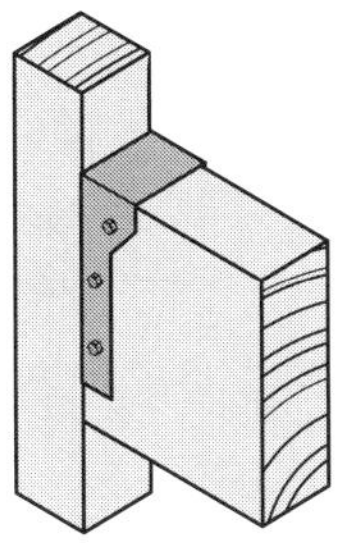

(A) The connector is improperly installed.

(B) The connector used is improper for the wood-to-wood connection.

(C) The connector is installed correctly.

(D) none of the above

54. Construction detail sheets normally communicate information through

(A) dimensions

(B) isometric drawings

(C) materials

(D) orthographic drawings

55. Contractor's shop drawings are transmitted to the landscape architect for approval as part of the project

(A) plans

(B) submittals

(C) performance bond

(D) pay applications

56. In the concrete curbing detail shown, which element serves as the concrete reinforcement?

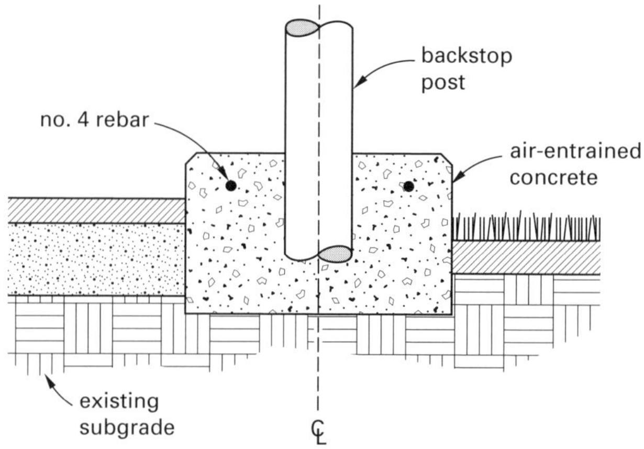

(A) no. 4 rebar
(B) backstop post
(C) existing subgrade
(D) air-entrained concrete

57. A legal description of a parcel would be found on the cover sheet of a construction document set and contain what information?

(A) benchmark elevation
(B) fixed location of the parcel
(C) monument legend
(D) surveyor of record

58. When setting the base elevation in a survey, what should be used to relate the information to the adjoining properties?

(A) corner pin
(B) benchmark
(C) hub and whiskers
(D) lot-line bearing

59. What critical component is NOT shown in the given wall detail?

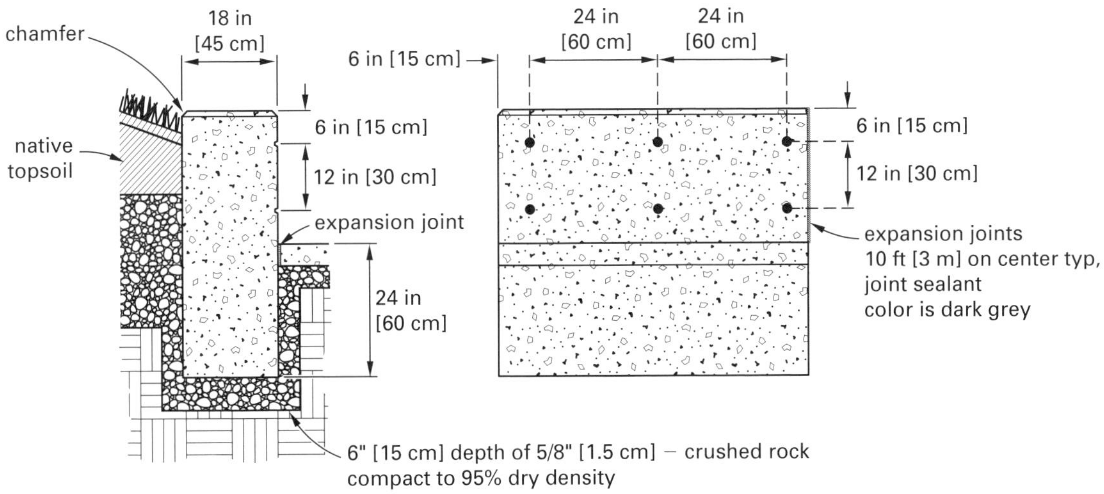

(A) reinforcement
(B) notes on the finish
(C) dimensions
(D) subgrade preparation

Construction Documentation

60. Construction on a project is governed by the contract documents. What items comprise a legal set of contract documents?

I. working drawings
II. specifications
III. general conditions
IV. form of agreement

(A) I and II
(B) I and III
(C) II and IV
(D) all of the above

61. The cover sheet on a set of construction plans normally includes which of the following information?

I. client names
II. consultant names
III. property legal description
IV. vicinity map
V. project name

(A) I and II
(B) II and V
(C) III and IV
(D) all of the above

62. When writing a legal contract with an owner for construction phase services, which of the following terms should be avoided in reference to the role of the landscape architect during the construction phase?

I. landscape architect
II. supervision
III. observation
IV. inspection

(A) I
(B) III
(C) II and IV
(D) III and IV

63. The primary role of regulatory agencies during the construction phase of a landscaping project is to

(A) verify compliance with codes, ordinances, and regulations
(B) provide oversight on behalf of the landscape architect
(C) collect permit fees
(D) collect information on the contractor's license

64. When bidding a project, the bidding contractor's primary sources of information are

(A) site visits
(B) addenda
(C) specifications
(D) construction documents

65. In the Construction Specifications Institute's (CSI) MasterFormat™ 2004 Edition, the irrigation section is located in

(A) Division 2
(B) Division 11
(C) Division 32
(D) Division 33

66. In the specifications for a project, documents relating to contracts, work sequences, price and payment procedures, and quality requirements should be placed in

(A) each specification section
(B) the form of bid
(C) the Division 1 general requirements
(D) the architect's supplemental information

67. When purchasing water rights, what formula is used to calculate the total amount of irrigation water needed?

(A) maximum applied water allowance
(B) total system pressure loss
(C) total dynamic head
(D) total gallons per minute (gpm) [liters per minute (L/min)]

68. The role of the landscape architect during construction is defined during the

(A) bidding phase
(B) construction document phase
(C) contract negotiations with the owner
(D) permit phase

69. For a construction project taking place in wet or freezing conditions, what is a proper base product to use under the asphalt wearing course?

(A) colloidal soil
(B) asphalt treated base
(C) concrete
(D) nonwoven geotextile fabric

70. If a specified construction product is not available, or if the contractor would like to use a different product, the landscape architect should review and approve the new product choice through

(A) a pay application
(B) performance bonds
(C) a substitution request
(D) specifications

71. An underground irrigation system is to be installed in a colloidal soil. Which piece of equipment could be used to accomplish this task in the most economical way?

(A) chain trencher
(B) trackhoe
(C) box scraper
(D) dozer

72. Connecting an irrigation system to a potable water source requires the use of

(A) a pressure regulator
(B) a backflow prevention device
(C) an electric valve
(D) irrigation heads

73. In the United States, soil is tested and evaluated for planting conditions using the

(A) AASHTO system
(B) ASTM system
(C) USDA system
(D) unified system

74. The proper sequencing of a construction project is

I. receiving bids and awarding the project
II. installing irrigation and plants
III. performing mass earthwork
IV. installing minor site structures

(A) I, II, III, IV
(B) I, III, IV, II
(C) II, IV, I, III
(D) III, I, IV, II

75. On a road improvement project, the nearest quarry is 50 mi [80 km] away. On-site space is not an issue. Which of the following is the LEAST expensive method of preparing a road base?

(A) Import structural fill materials.
(B) Grind, stockpile, and reuse the existing asphalt section.
(C) Import clean, to-specification crushed aggregate.
(D) Install a lift of asphalt treated base.

76. Which piece of equipment would most economically complete the task of moving 10,000 yd^3 [7645 m^3] of materials on site?

(A) bobcat
(B) box scraper
(C) dozer
(D) backhoe

77. The most economical construction staking method for a large-scale state highway construction project is to use

(A) a global positioning system unit
(B) a laser transit
(C) a survey crew
(D) links, or a Philadelphia rod

78. When installing a drainage system of 4 in to 6 in [10 cm to 15 cm] perforated pipe in a sandy loam golf course, which piece of equipment would be the MOST economical to use and cause the LEAST amount of damage to the site?

(A) backhoe
(B) trackhoe
(C) turf tire wheel trencher
(D) irrigation shovel

79. Which of the following is the LEAST expensive fastener type for use in securing a wood deck member to a wood beam?

(A) nail
(B) screw
(C) lag bolt
(D) anchor bolt

80. Construction surveying on a project is critical for locating building corners, connections to utilities, grading, and other site amenities. What coordinate type is used to locate these items?

(A) northings and eastings
(B) azimuth
(C) Cartesian
(D) meridian

81. In the calculation of a bearing, what should serve as the starting point?

(A) east or west
(B) basis of bearing
(C) north or south
(D) the transverse

82. In the construction documents, the requirements for stripping and stockpiling existing topsoil for planting would typically be found in the

(A) general conditions
(B) Division 31 specifications
(C) planting plan
(D) contract closeout documents

83. The primary job of the landscape architect during the construction observation phase of a project is to

(A) review contract labor laws
(B) instruct on methods of construction
(C) ensure compliance with construction documents
(D) address change orders

84. The best means of ensuring quality control during a construction project is to

(A) process the submittals
(B) provide construction observation
(C) process the pay applications
(D) conduct phone conversations with the contractor

85. For some aspects of a project, the contractor may be given latitude in deciding methods of construction. Prior to final acceptance and final payment to a contractor, these items should be provided to the landscape architect, for approval, in the

(A) submittals
(B) inspection reports
(C) permit
(D) as-built drawings

86. At what point during the construction phase of a project are survey control points typically installed?

(A) prior to award of the bid
(B) after clearing and grubbing
(C) during paving
(D) prior to site clearing

Materials and Methods of Construction

87. When designing and installing a landscape timber wall with 4" × 6" treated materials four pieces high, what type of fastener is most appropriate to connect the timber layers?

(A) spike
(B) wood screw
(C) cavity wall tie
(D) snap tie

88. Which type of application would typically need an approved solid, structural fill material for construction?

(A) floor slabs and building foundations
(B) a natural grass ballfield
(C) wetland restoration
(D) a chain-link fence

89. Number 4 steel rebar has a diameter of

(A) $1/4$ in [6 mm]
(B) $1/2$ in [1.25 cm]
(C) $5/8$ in [1.5 cm]
(D) 1 in [2.5 cm]

90. What association sets the standards for concrete in the United States?

(A) DIN
(B) TUV
(C) ASTM
(D) CSA

91. Which surveying method takes into account the curvature of the earth and would be used on large-scale planning projects?

(A) metes and bounds
(B) plane surveying
(C) geodetic surveying
(D) global positioning

92. What are the primary considerations in designing a wooden deck?

I. framing
II. materials
III. maintenance
IV. staining

(A) I and II
(B) II and IV
(C) I, II, and III
(D) I, III, and IV

93. Wood members for a deck should be sized according to

(A) ASTM standards
(B) connector size tables
(C) wood span tables
(D) sieve analysis

94. Live loads designed for a wooden deck are typically measured in

(A) pounds per square foot (psf) [kilograms per square meter (kg/m^2)]
(B) pounds (lbs) [kilograms (kg)]
(C) pounds per square inch (psi) [kilograms per square centimeter (kg/cm^2)]
(D) feet (ft) [meters (m)]

95. What unit is used to measure illumination on a surface?

(A) lumen [luminaire]
(B) candelas per square inch [candelas per square meter]
(C) footcandle [lux]
(D) British thermal unit [joule]

96. During the installation of a large irrigation system, irrigation sleeves should be installed after

(A) earthwork is completed
(B) site concrete is poured
(C) paving is completed
(D) planting is completed

97. When mixing irrigation heads of various radius sizes, the irrigation head nozzles must have matching

(A) flow rates
(B) precipitation rates
(C) inlet sizes
(D) velocities

98. When designing an irrigation system, the valve wire used must be sized to allow for the voltage drop at the end of each run of wire. The voltage drop is determined using

(A) Ohm's law
(B) the Bernoulli effect
(C) the pressure loss coefficient
(D) the modified Proctor test

99. The relative compressive strength of a typical concrete sidewalk is directly related to the ratio of what two products?

I. water
II. cement
III. aggregate
IV. reinforcing material

(A) I and III
(B) I and IV
(C) II and III
(D) III and IV

100. The full strength of a standard concrete mix is reached after

(A) 1 day
(B) 2 days
(C) 12 days
(D) 28 days

101. What chemical process takes place with the mixing of cement and water?

(A) entrainment
(B) hydration
(C) oxidation
(D) abrasion

102. What type of admixture is used to slow down the setting of concrete?

(A) reducer

(B) air entrainment

(C) accelerator

(D) retardant

103. What are the most common masonry materials used in design and construction projects?

I. clay
II. brick
III. concrete
IV. stone

(A) I and II

(B) II and III

(C) I, III, and IV

(D) II, III, and IV

104. When installed in a landscaping project, the benefits of masonry products over other product types include greater

I. weight
II. durability
III. compressive strength
IV. moisture resistance

(A) I and II

(B) III and IV

(C) I, III, and IV

(D) all of the above

105. What makes a solid, watertight masonry joint?

(A) compressed, tooled joints

(B) contraction joints

(C) protruding mortar joints

(D) expansion joints

106. What basic functions does mortar provide in masonry unit construction?

I. bonding units
II. compensating for irregularities in production of units
III. allowing use of ties and reinforcing integral to the unit
IV. enhancing aesthetics through color and texture

(A) I and II

(B) I and III

(C) II, III, and IV

(D) all of the above

107. In a masonry unit wall, what type of reinforcement would be used to hold veneered clay brick units?

(A) steel rebar
(B) wall-tie tabs
(C) anchors
(D) reinforcing wire

108. What product should be used to install a handrail in a sleeve embedded into a concrete masonry unit wall?

(A) epoxy
(B) non-shrink grout
(C) mortar
(D) plate anchor

109. The tensile strength of a wood member refers to the product's resistance to

(A) loads that cause the member to twist
(B) loads that cause the member to compress
(C) stresses parallel to the wood grain
(D) loads perpendicular to the wood grain

110. Nominal-sized lumber is sized prior to

I. kiln drying
II. shrinkage
III. planning
IV. installation

(A) I
(B) I and II
(C) II and III
(D) II and IV

111. A wooden ledger is to be designed for a new deck attached to a new concrete wall. Which fastener type is most appropriate for this situation?

(A) machine screw
(B) lag bolt
(C) common nail
(D) anchor bolt

112. What factor will cause a construction material, such as soil or aggregates, to reduce in volume?

(A) tension
(B) compaction rate
(C) compressive force
(D) impact velocity

113. When the dead load design weight of a deck is too great, the deck posts will

(A) bend
(B) deflect
(C) be crushed
(D) twist

114. The major load-carrying member in a wooden deck design is the

(A) decking
(B) beams
(C) joists
(D) rails

115. Beams that span more than two posts are called

(A) cantilevers
(B) continuous
(C) overhangs
(D) loadbearing

116. What is the difference between geodetic surveying and plane surveying?

(A) The distance between control points is different.
(B) A plane survey uses the global positioning system.
(C) A geodetic survey uses a transit.
(D) A geodetic survey takes into account the earth's curvature.

117. In plane surveying, distances are most commonly measured

(A) algebraically
(B) along the horizontal plane
(C) with a foresight and backsight
(D) with a benchmark

118. What key soil characteristic should be factored into the sizing of a footing?

(A) soil classification
(B) soil structure
(C) bearing capacity
(D) soil density

119. What three main components make up a soil?

(A) soil, bark, mulch
(B) silt, sand, water
(C) silt, sand, clay
(D) aggregate, soil, mulch

120. The angle of repose is the

(A) best angle at which to hydroseed a soil
(B) natural angle at which a soil rests
(C) angle from which a soil will compact
(D) compacted, preloaded angle of a soil

121. Which pH listed is closest to neutral?

(A) 0
(B) 5
(C) 6
(D) 9

122. What soil characteristic is important when designing a very heavily loaded footing?

(A) liquefaction
(B) permeability
(C) plasticity
(D) cohesion

123. At what point in time after a major moisture event is the plastic limit of a soil achieved?

(A) 24 hours after the moisture event stops
(B) when enough moisture leaves the soil profile and the soil is no longer in a liquid state
(C) when the soil cohesion fails
(D) when the soil moisture content reaches saturation

124. Which wall type is NOT a unit, or stack, retaining-wall system?

(A) dry-stack stone
(B) gabion
(C) cantilever
(D) crib and bin

125. Which of these classification types is NOT a form of retaining wall?

(A) sill and ledger
(B) reinforced embankment
(C) unit stack
(D) cast in place

126. In which type of wall is a weep hole appropriate to use as the form of drainage?

(A) any rigid wall
(B) rockery construction

(C) a precast unit wall
(D) a gabion unit wall

127. The weight of the soil being held back against a retaining wall contributes what factor to the wall?

(A) batter
(B) stability
(C) lateral load
(D) strength

128. On a job site, the general contractor's most important responsibility is to

(A) provide a project engineer
(B) complete the work in a safe, technically competent, and efficient manner
(C) work at a pace he or she feels is appropriate
(D) review documents for change orders

129. The base aggregate layer of an asphalt roadway should extend beyond the finish pavement edge to

(A) provide an adequate shoulder for driving safety
(B) satisfy municipal code requirements
(C) improve drainage and pavement support
(D) provide a planting strip

130. Vibration is used in a concrete pour to

(A) liquefy the concrete
(B) consolidate the concrete and fill voids in forms
(C) bring the fine particles to the surface
(D) provide for a quality finish

131. The most common type of concrete placement is

(A) cast-in-place
(B) mixed-in-place
(C) sprayed or air blown
(D) precast

132. What aggregate material is NOT commonly used in a concrete mixture?

(A) sand
(B) crushed stone
(C) gravel
(D) cobblestone

133. Which of the following is a typical concrete test method?

(A) clump test
(B) pressure-loss test
(C) slump test
(D) modified Proctor test

134. In preparing a site for the installation of concrete, a landscape architect should ensure that the subgrade meets the specifications for preparation by using

(A) the cylinder test
(B) the Proctor test
(C) a nuclear densometer
(D) the slump test

135. Which surveying method is typically used on small to medium landscape architecture projects?

(A) metes and bounds
(B) plane surveying
(C) geodetic surveying
(D) reference datum

136. What components are combined to make mortar?

I. water
II. gravel
III. cement
IV. sand

(A) I and II
(B) II and III
(C) I, III, and IV
(D) all of the above

137. The term "unseasoned" is used in reference to what kind of lumber?

(A) freshly cut
(B) kiln-dried
(C) air-dried
(D) common

138. Finish-grade lumber is usually specified as having a kiln-dried appearance. What mark appears on the lumber adjacent to the "kiln dried" stamp?

(A) percent of pitch
(B) timber lot
(C) moisture content
(D) indication of the finished side

139. Lumber in North America is sold in units of

(A) square feet
(B) cubic yards
(C) board feet
(D) board length

140. When a project consists of multiple irregular parcels that the owner wants to consolidate, what type of survey should be done?

(A) ALTA survey
(B) topographic survey
(C) GPS mapping
(D) boundary line adjustment

141. A connection must be made in a deck construction project that requires both bonding wood to wood and using a sealant to protect the deck from moisture. What fastener could be used in conjunction with the bolting system?

(A) wood glue
(B) mortar
(C) construction adhesive
(D) bituminous binder

142. When compacting a colloidal clay soil, what type of compactor is best suited for an area covering 2.5 ac [1 ha]?

(A) plate compactor
(B) double-drum vibrating roller
(C) sheep's foot roller
(D) walk-behind tamper

143. What product would be specified for installation between two lifts of new asphalt paving material?

(A) primer
(B) tack coat
(C) binder
(D) epoxy

144. Welded wire mesh used in a concrete slab serves as a

(A) construction joint
(B) means of increasing slab tensile strength
(C) concrete additive
(D) contraction joint

145. What is the minimum depth requirement for a contraction joint?

(A) 1/2 in [12.7 mm]
(B) the depth of a concrete trowel
(C) 1/6 the depth of the slab
(D) the maximum size of the aggregate

146. When concrete is to be subjected to roadway conditions, the primary criterion for selecting the aggregate used in the concrete is

(A) rock hardness
(B) quantity of aggregate in the mix
(C) range of aggregate size
(D) local availability

147. Materials added to concrete before or during the mixing process are called

(A) aggregates
(B) water
(C) admixtures
(D) cement

148. In the process of rough grading, a soil is found to have a high moisture content and a potential for liquefaction. Which equipment type should be specified for construction?

(A) wheeled equipment
(B) a skid-steer loader
(C) drag-line buckets
(D) low-ground-pressure, tracked equipment

149. The recommended length for nails specified for the installation of wood decking is

(A) the same as the length of the joist hanger being used
(B) 1 1/2 times the size of the wood member being secured
(C) 2 1/2 times the thickness of the wood member being secured
(D) long enough so that half the nail is embedded into the framing material

150. In designing a footing for a deck post, the soil is found to have a bearing capacity of 7 psi [5000 kg/m^2], and the design load is 4000 lbf [1800 kg]. What minimum square footing size is required?

(A) 1/2 ft × 1/2 ft [0.1 m x 0.1 m]
(B) 1 ft × 1 ft [0.3 m x 0.3 m]
(C) 2 ft × 2 ft [0.6 m x 0.6 m]
(D) 3 ft × 3 ft [0.9 m x 0.9 m]

151. What is the proper orientation of wood decking material?

(A) flat grain down
(B) alternating vertical grain and flat grain
(C) bark side up
(D) bark side down

152. In the installation of an exterior landscape lighting-control panel, how should the system be grounded?

(A) Drive a piece of rebar into the ground.
(B) Install a ground wire with a welded connection.
(C) Ground the panel to the adjacent metal downspout.
(D) Ground the system to the metal conduit with incoming wires.

153. What is the best type of soil to specify as a structural fill material?

(A) silts
(B) crushed aggregates
(C) graded gravels
(D) graded sands

154. What does DBH stand for?

(A) diameter base height
(B) diameter breast height
(C) depth of bury height
(D) diameter branch height

155. What type of wall is made from wire baskets and filled with crushed rock?

(A) rockery wall
(B) cantilever wall
(C) gabion wall
(D) gravity wall

156. The shallow angle constructed into the face of a retaining wall is known as the

(A) slope
(B) batter
(C) tip
(D) slant

157. Which of the following statements applies to the rain garden detail shown?

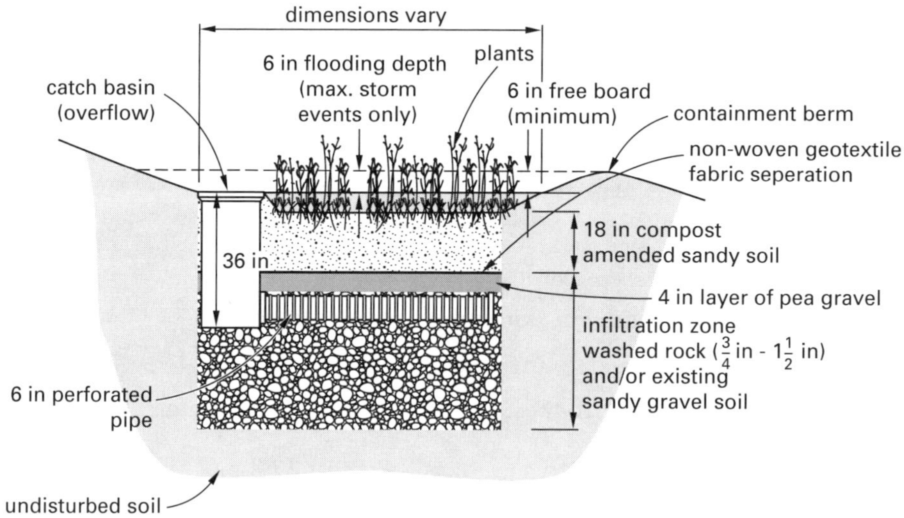

cross section view
(not to scale)

(A) The detail is correct as is.

(B) The perforated pipe is not in the correct location.

(C) The free board area is not of proper depth.

(D) The geotextile fabric is inappropriate.

Solutions

Design Principles

1. In irrigation system design, pipe is sized based upon the velocity of water that will flow through it, in feet per second [meters per second]. Velocity is the measure of distance traveled over a period of time. A pipe's diameter and material will affect the velocity of the water flowing through it. When designing an irrigation system, the landscape architect should refer to velocity charts and select a pipe that supports the project requirements.

The answer is (B).

2. Sandstone is a good masonry selection for a patio project that is exposed to the elements. Sandstone is fairly smooth yet has a nonslip surface. And it comes in large pieces, yet small joints can be created between pieces.

Marble would be too smooth and slick when wet or frozen, as would smooth granite. Although thermal finish granite could be used, this production procedure is not standard for granite and cannot be assumed from the options given. Cobblestone is durable, but is too bumpy and irregular for a patio environment.

The answer is (C).

3. Deck and platform railing design codes require the spacing between rails to be no more than 4 in [10 cm]. This requirement is based on the size of a baby's head.

The answer is (C).

4. The Occupational Safety and Health Administration (OSHA) requires deck and platform railing design codes to be at least 42 in [107 cm] high.

The answer is (D).

5. The goal of landscape lighting is first and foremost to provide safety for the user, second to encourage the use of the space at night, and third to provide legibility of nodes (i.e., focal points) for people using the space at night. The use of lighting elements will help people feel safe and comfortable and will create a space whose usefulness extends beyond daylight hours.

The answer is (D).

6. In a recreational area, uniformity is the most critical factor to design for. Non-uniform lighting on a sports field results in shadows and blind spots that can reduce safety and lead to injury.

The answer is (B).

7. The goal of installing pathway lighting is to provide a safe environment for people to travel through at night. Types of pathway fixtures and their frequency of placement can depend upon other surrounding elements and upon the lighting level desired within the landscape.

The answer is (B).

8. While there are various physical properties to evaluate with each soil, the terms "physical," "site engineering," and "horticultural" will cover all aspects of soil evaluation.

The terms "agricultural" and "horticultural" mean the same thing. The terms "colloidal" and "textural" refer to actual physical properties that must be included with others for proper evaluation, which is often performed by a geotechnical engineer.

The answer is (A).

9. In an urban environment, soil will be walked on, driven on, and generally poorly maintained, resulting in the loss of air voids for oxygen exchange in the soil. Compaction results in lower infiltration rates, and lack of maintenance results in loss of organics. Or, too often, all the organic debris—such as leaves and grass clippings—is removed during the maintenance process. In recent years, the study of mixing organics with fine aggregates has created structural soils that have been successfully installed in many urban landscapes, resulting in healthy plantings.

The answer is (A).

10. The long-term success of any planting plan depends primarily on the availability of maintenance dollars from the owner. If the maintenance budget is limited, plant types and installation methods become the key contributors to long plant life.

The answer is (A).

11. The United States Department of Agriculture's (USDA's) hardiness zone map gives a general indication of what species will grow in what climatic conditions around the country. However, the survival of each cultivar of a specific species may depend upon where the plant is grown and what nursery it came from. The landscape architect should check with a local, reputable nursery to verify proper plant cultivars for the particular region.

The answer is (B).

12. All walls should be designed with some form of drainage. The drain may consist of weep holes through a cast-in-place wall, angular backfill, or just natural voids in the rocks or precast units. But all walls need some means of releasing water from the retained soil to avoid structural failure.

The answer is (B).

13. During the planting of ball-and-burlap materials, the only pruning that should take place is the removal of dead or damaged braches. Removing any other growth will not aid the establishment of ball-and-burlap plants. Native species, however, may flourish with pruning after planting.

The answer is (B).

14. Zinc is the coating applied to nails used in exterior applications. Zinc is applied to nails in a process known as galvanization.

Aluminum is not a nail coating. Steel, if left untreated, would rust and stain the wood product. Copper is used for piping, but it also oxidizes and would stain wood products if used as a nail coating.

The answer is (B).

15. Retaining walls can be designed up to 3 ft [1 m] without the aid of a structural engineer in most states and provinces. When a wall exceeds a height of 3 ft [1 m], consulting a structural or civil engineer will ensure the health and safety of the site's users, avoiding the potential for a wall collapse.

The answer is (C).

16. The height of a retaining wall is measured from the bottom of the footing—including rocks and concrete units buried below the lower finish grade of the soil—to the top of the wall. Running design calculations on measurements that are not based on this starting point could cause a failure. Secondarily, measuring only from the lower finish grade will result in inaccurate cost estimates, because even the buried portions of the wall and footings represent a contractor's expense that should be passed on to the owner.

The answer is (D).

17. Pyramidal evergreen shrubs would prevent a vehicle's headlight glare from leaving the site. Neither low flowering shrubs nor tall canopy shade trees are dense enough to block the glare from leaving the site.

The answer is (B).

18. To allow for the safe entrance and exit of vehicles from a site, landscape architects must establish a site triangle (an area where all plantings must be under 30 in [76 cm], so that drivers have clear visibility of oncoming traffic and pedestrians). The dimensions of the site triangle depend on the speed of the road. Drivers need to see farther when cars are driving faster, so the site triangle must be larger. In this problem, the speed was used to indicate that this was a major road, so the site triangle would be correspondingly large. Planting design for sidewalk planters, parking islands, and parkway planters depends upon many other factors.

The answer is (C).

19. Adjacent to any entry doors that will be used after dark, dense plants, such as evergreen shrubs or trees, should not be planted. Evergreen shrubs or trees limit visibility, making it difficult for security people or cameras or see through them. They also provide enough coverage for pedestrians to hide. In order to promote safety, the bank should instead plant canopy trees, ground covers, or low flowering deciduous shrubs.

The answer is (A).

20. A south-facing, central courtyard plaza will receive direct sunlight and reflective heat from the hardscape. Only canopy trees would create enough shade to allow for the comfortable use of the courtyard area during mid-day lunches and/or performances.

The answer is (C).

21. The source of the landscape's lighting should not be visible. If viewed directly, a fixture's lightbulb can cause a "blinding effect" depending on its type and intensity. With the proper use of a fixture's location, viewers can enjoy the effects of a site's lighting design and its enhancement of the landscape.

The answer is (C).

22. Low-voltage lighting can be installed by most licensed landscape contactors with limited permit requirements. High-voltage and sports-field lighting both typically require the issuance of an electrical permit, may required review by a jurisdictional agency, and must be installed by a licensed electrical contractor.

The answer is (A).

Resource Conservation and Management

23. In applications that have poor soil conditions in the subgrade, it is important to install a separation barrier between the subgrade and basecourse soils. This keeps the subgrade from migrating upward, and keeps the base materials from migrating or being lost into the subgrade. The most common and cost-efficient means of achieving this is to install a geotextile fabric between the subgrade and base course material.

The answer is (B).

24. Fall is the best time to plant the majority of materials. Planting is itself a stress factor, which can be worsened by heat, drought, frost, or other severe weather conditions. Fall planting allows plants to establish root growth prior to encountering these seasonal, weather-related stresses.

Spring planting is preferable to summer or winter planting, but it does not allow for good root growth—only top growth flushing. In the spring, plant growth is focused on creating the sugars needed for rapid and abundant leaf and top growth, and for flowering. To force a plant to grow roots in the spring, heavy pruning must be done, but this can compromise the intent of selecting that plant for the location.

The answer is (D).

25. The American Nursery and Landscape Association (ANLA) (formerly the American Association of Nurserymen) is the governing body recognized as setting the standards for the health, size, shape, and vigor of plant materials. While other such associations exist, they govern other areas of projects, such as earthwork, compaction, and specification writing.

The other abbreviations listed stand for American National Standards Institute (ANSI), American Society for Testing and Materials (ASTM), and Construction Specifications Institute (CSI).

The answer is (C).

26. Bioengineering in landscape architecture involves the establishment of plant materials on hillsides, or on the banks of streams, canals, or ponds to control erosion and prevent excessive runoff. A successful bioengineering project will use live stakes and small plants and will involve collaboration with other design professionals.

The answer is (C).

27. Acrylic latex paint is a water-based product and would be best suited to an exterior environment that is not exposed to weather.

Oil-based, or polyurethane, products—such as transparent stain or alkyd paint—are appropriate for use in an exterior exposed installation condition.

The answer is (A).

28. The frost line, or frost depth, must always be considered when designing in a cold climate to determine the underground depth that will protect irrigation pipes from a freeze.

Bearing capacity is a function of the soil, water, and footing or foundation, and how they interact with each other (i.e., slip forces, downward pressures, compaction around the footing, etc.). Soil classification and growth seasons dictate types of plant materials selected.

The answer is (C).

29. The best way to prepare a planting hole is to make the hole shallow and wide. This allows sufficient air and moisture to get to the roots of a newly planted tree or shrub, promoting growth. Avoid preparing deep, long pits for plants, as this compacts the soil and does not allow for free, unrestricted root growth.

The answer is (B).

30. For ball-and-burlap trees, the twine, basket, and burlap should be fully removed from the pit—if at all possible without damaging the root ball during the removal process. These materials should not be folded down into the planting pit. Doing so would leave air voids and gaps that would impede root growth and damage the tree's health, since the wire, twine, and burlap will not decompose quickly.

The answer is (C).

31. A 20 in [50.8 cm] frost depth should be taken into account when trenching for a mainline. The trench illustrated is 24 in [61 cm], so the irrigation mainline shown can have a diameter of at most 4 in [10.2 cm]. Of the options shown, *C* (3 in [7.5 cm]) is the only suitable choice. This allows for 21 in [53.5 cm] of cover material over the mainline, which is sufficient to protect the pipe from frost. The larger the pipe is, the deeper the trench needs to be to ensure proper coverage. The type of traffic that will occur over the pipe should also be considered.

The answer is (C).

32. Ammoniacal copper arsenate (ACA), chromated copper arsenate (CCA), and wolmanization are three types of lumber treatments for outdoor applications that involve ground contact. These treatments keep the lumber from rotting and often contain poisons to kill termites and other insects that try to eat the wood. Most treated wood has a life expectancy of 20–25 years.

The answer is (D).

33. 4 ft [1.2 m] is the maximum depth to stockpile soil. Deeper piles compact under their own weight, removing oxygen and thereby killing off important soil organics. Shallower piles can be used, but they take up large areas.

The answer is (C).

34. The installation of a construction-barrier fence at the dripline will keep the topsoil from compacting and allow moisture to soak into the soil, preventing physical damage to tree roots. Provision for monetary damages should be included in the contract documents to allow for recovery and replacement by the owner if items are damaged by the contractor.

The answer is (B).

35. In the installation process, the single most important contributor to long-term plant health is the condition of the plants upon arrival at the planting site. Plants should be inspected for dryness and damage, and trees should be examined for unhealed wounds in the bark.

The answer is (C).

36. Burying the stem or trunk of a plant is not beneficial, as it will only rot the stem. Mulch should be kept away from the trunk or stem and spread completely over the plant's root zone.

The answer is (C).

37. Staking should only occur where called for by severe weather conditions, such as high winds, or when the size of the tree does not allow for self-support. When used, stakes and guy wires should be loosened after the first growing season and removed after the second growing season.

The answer is (C).

38. Low Impact Development (LID) is an alternative, comprehensive stormwater management strategy focused on the reduction of environmental and economic impacts of site development. LID uses sustainable materials and concepts to maintain the natural hydrologic cycles that existed on-site prior to any development. LID is especially appropriate for urban redevelopment projects and/or for projects associated with existing environmentally sensitive areas.

The answer is (B).

39. LEED (Leadership in Energy and Environmental Design) is a program developed by the U.S. Green Building Council (USGBC) to measure the performance efficiency and sustainability of buildings and their associated development. LEED programs include site development and the reuse of resources, placing important emphasis on the reduction and reuse of water for irrigation. LID (Low Impact Development) focuses on managing water and stormwater runoff to help protect a site's natural hydrologic regimes. BMPs (Best Management Practices) govern not only stormwater, but also irrigation and on-site landscape management issues, through the use of native plantings; reduced use of fertilizer, pesticides, and herbicides; and the reuse and proper amendment of on-site soil.

The answer is (D).

40. The benefits of Low Impact Development (LID) extend far beyond sustainable stormwater management. In addition to managing stormwater runoff, the small-scale rain gardens (also known as bioretention systems) used in LID designs can enhance landscape aesthetics, as well as provide additional spaces for wildlife forage and cover. LID designs reduce the need for sub-surface piping and associated detention pond infrastructures, offering potential cost savings and increases in available on-site total square footage.

The answer is (D).

41. Traditional site development and/or construction alter a site's existing hydrologic regime by decreasing soil porosity and infiltration capacity; increasing areas of impervious surfaces; and increasing the velocity, volume, and duration of stormwater runoff. Low Impact Development (LID) modifies traditional site development and construction practices to create a post-development site that mimics, as best as possible, the natural hydrologic regimes found on the pre-developed site.

The answer is (B).

42. For Low Impact Development (LID) projects, a landscape architect would employ the use of rain gardens (bioretention systems) and porous pavements. Rain gardens are point-source collection basins that collect, manage, and filter stormwater through the use of appropriate soils and vegetation. Porous pavements reduce overall impervious surfacing and increase stormwater infiltration capacity.

Discharge structures and detention facilities would significantly alter the hydrology of a project site because their traditional "pipe-and-pond" techniques do not allow rainwater to recharge into the ground through soil contact.

(Note: A landscape architect should maintain active communication with the project's civil engineer throughout a site's development to ensure the proper and successful integration of LID site design techniques, such as rain gardens or porous pavements.)

The answer is (A).

43. Rain gardens (bioretention systems) collect and store stormwater from impervious surfaces to limit excessive runoff and to recharge groundwater. Because a rain garden's multiple soil layers naturally filter pollutants often carried with runoff, rain gardens also improve the water quality delivered to existing soils, storm sewer systems, nearby streams, or lakes.

The answer is (D).

44. On-site soil reuse, also known as on-site recycling, has immediate and far-reaching environmental impacts. Regionally and globally, the reuse of on-site soils reduces fuel consumption and associated emissions by eliminating off-site hauling. It also reduces the unnecessary wasting of soil.

Locally, the reuse of on-site topsoil provides a site-specific soil mix. The process of creating a site-specific soil mix includes stockpiling the salvaged soils in proper windrows, tilling and amending the existing subgrades, and amending the stockpiled soils with certified compost material. Certified compost material is typically comprised of composted yard waste and/or site grindings from limbs, brush, and stumps that are removed during the site's development process.

The answer is (C).

45. One of the main goals of the Low Impact Development (LID) Best Management Practices (BMPs) for stormwater management is to recharge groundwater. Instead of using conveyance to remove stormwater from a site via sub-surface drainage pipes, LID BMP strategies promote on-site, point-source water management and mitigation by allowing stormwater to infiltrate native subsoils as if the site was undeveloped.

The answer is (C).

46. To follow stormwater Best Management Practices (BMPs), the rain garden's soil profile would consist of a compost-amended sandy soil. Compost-amended sandy soil allows water to infiltrate the rain garden's subsoils. It also increases the fertility and functional viability of the soil for both plant and microbial vigor, improves water quality, and retains water within the soil profile. The soil's ability to retain large volumes of water keeps moisture available to the rain garden's vegetation, and allows for reduced irrigation to the planted area.

The answer is (C).

47. Native vegetations (plants, shrubs, and trees that existed in an area before it was developed) are most appropriate for rain gardens without supplemental irrigation. Native vegetation is acclimated to the local climate and will adapt best to periods of high or low water availability.

Many trees, ornamental shrubs, and flowering plants require supplemental irrigation to survive dry periods. They may also have temperamental root systems that are not equipped to handle the moisture fluctuations in excessively wet or dry seasons.

The answer is (B).

Graphic Communication

48. Point A indicated on the given soil classification chart represents a soil that is 70% sand, 5% clay, and 25% silt. The United States Department of Agriculture (USDA) soil classification chart is based upon the agricultural suitability of soil. Knowing how to use this chart is critical for the many soil strategies that deal with stormwater, low-impact development, and drought conditions.

The answer is (D).

49. The fastener shown is a lag bolt, a fastener used to secure items to wood products.

A wood screw is used to attach decking or wood products in applications such as arbors or gazebos. A machine screw is used for metal-to-metal connections. Anchor bolts are typically set into concrete.

The answer is (A).

50. The fastener shown is a machine screw, with standard (even) threads. A machine screw is used to connect metal, plastic, or small wood pieces.

Lag bolts and wood screws are used exclusively for wood connections and have aggressive, wide thread patterns.

The answer is (D).

51. The fastener shown is a wood screw. Wood screws are used to hold small wooden pieces, decking, and nonlateral sheer structural elements. Small wood screws will shear in a lateral direction, so the use of wood screws is prohibited by code in many situations.

The answer is (A).

52. The fastener shown is an anchor bolt and would be used to attach a concrete footing to a landscape element. Anchor bolts are cast in place to provide the structural tie to the footing for landscape elements such as benches, light standards, and walls.

An anchor bolt is not proper to use for a roof truss or to attach new wood structures to existing wood structures or beams.

The answer is (A).

53. The connector shown is installed upside down and will not carry and transfer the load from the joist or beam to the post.

The answer is (A).

54. When conveying information in construction details, all aspects of a design must be shown in a two-dimensional plane—so multiple views, angles, and sections are used to convey the intent. This multi-perspective drawing method is known as orthographic drawing.

The answer is (D).

55. When a contractor provides the landscape architect with information on a shop drawing, or with product data, it is known as a submittal. The proper procedure (timelines, paperwork, etc.) for conveying submittals is contained within the project specifications (typically in Division 1).

The answer is (B).

56. In the concrete curbing detail shown, the number 4 rebar is the reinforcement method used.

A backstop post is an element that provides no reinforcement of a running curb. The existing subgrade is the prepared base on which the element is constructed; it provides no structural reinforcement to the curbing. Air entrainment is an additive that produces smoother, longer lasting concrete; it does not reinforce the concrete.

The answer is (A).

57. The cover sheet in a set of construction documents is where the fixed location, or legal description, of the parcel will be located. Legal descriptions are a critical piece of data for the permit process, as most permitting agencies derive all their data from the legal description.

The answer is (B).

58. A benchmark is used to set elevations related to a landscape architecture project. All topographic surveys indicate the datum and benchmark that the survey is tied to for its elevational information. This benchmark is used as the base reference point in the designing of a project's grade elevations.

The answer is (B).

59. The given wall detail shows a seat wall that is holding back a hillside of soil. The soil will exert pressure against the back of the wall, but the detail does not show any reinforcement to counteract this pressure. The unreinforced wall could fail, due to lack of tensile strength, when pressure is exerted from the uphill force of the soil.

The answer is (A).

Construction Documentation

60. Contract documents include all of the items listed. Working drawings show the intent of the end product. Specifications are the written description of the drawings. The general conditions govern the how, when, and where, and other such issues associated with construction. And the form of agreement addresses the form of payment used between the owner and contractor and acknowledges the obligations of the contractor in the drawings and specifications.

The answer is (D).

61. The cover sheet should include all pertinent contact and site information, such as the vicinity map, contacts, permits, project name and address, parcel numbers, and items critical to identifying the project.

The answer is (D).

62. For the construction phase of a project, any reference to supervision or inspection has been dropped from the job description of the landscape architect. Supervision or inspection implies control over the contractor and can imply that the landscape architect should be responsible for the contractor's means and methods. Directing the contractor's means and methods would place the burden of the contractor's success or failure upon the landscape archtiect—a contractual liability that the landscape architect should avoid.

The answer is (C).

63. Regulatory agencies are solely responsible for ensuring compliance with all local codes, ordinances, and regulations applicable to the construction phase of a landscaping project.

The answer is (A).

64. During the bidding phase of a project, a contractor will use the construction documents—which include the drawings and specifications—to bid the project. Any addenda would become part of the construction documents.

The answer is (D).

65. MasterFormat™ 2004 Edition expanded on the 1995 edition to reflect new advances in the construction industry. While the 1995 edition places irrigation in Division 2, the MasterFormat 2004 covers irrigation in Division 32, which discusses external improvements. Division 11 covers equipment and Division 33 covers utilities.

A complete list of the MasterFormat™ 2004 Edition numbers and titles can be found on the Construction Specifications Institute's website, **www.csinet.org**.

The answer is (C).

66. Division 1 of MasterFormat™ 2004 Edition covers general requirements and should be used to organize documents such as contracts, work sequences, price and payment procedures, and quality requirements.

The form of bid will give the contractor's contact and licensure information, the bid amount, information on the contractor's bond, and other information required to evaluate the bid.

The answer is (C).

67. In a water-shortage situation, in order to properly estimate how much water will be used or how much water use will be allowed on a site, perform a calculation using the maximum applied water allowance (MAWA) system.

Capacity, pressure, and uniform distribution are all functions of the actual system and do not measure the amount of water the landscape architect will be allowed to use on a project.

The answer is (A).

68. The role of the landscape architect during construction is determined during the contract negotiation phase. Decisions made in this phase will affect how the contract documents are prepared, including which party is responsible for various areas of work during construction, such as submittal review, pay application review, construction administration, and legal paperwork as required by the contract documents.

The answer is (C).

69. For site construction in wet or freezing conditions, the design and use of an asphalt treated base (ATB) can allow for many areas of work to continue despite adverse weather conditions. An ATB can reduce subgrade damage caused by wet conditions. Further, it can reduce erosion (control points stay intact and in place) and reduce costs for storage and site clean-up.

The answer is (B).

70. When a landscape architect specifies a product that is no longer available and a substitution must be made, or when the contractor would like to use an equivalent or better product, these items are submitted to the landscape architect in the form of a substitution request. A substitution request should include all pertinent information necessary for the landscape architect to make a qualified comparison of the new product and the specified product. Information on how to submit a substitution request can be found in the project specifications.

The answer is (C).

71. There are many ways to install an irrigation system, but in all these methods the piping will go into a trench. Depending upon soil conditions, there are various ways to create the trench. A chain trencher will be the fastest, cleanest method for use in a colloidal, or cohesive, soil. Because chain trenchers are small pieces of equipment, their rental fees are relatively low and they can easily be transported from the rental yard to the site on a trailer. Additionally, they can be operated at a low labor rate, since most states permit them to be run by a landscape laborer rather than an equipment operator. A chain trencher will dig to the set depth and remove soil from the trench, and it can achieve several hundred feet of trenching in a single day with a single operator. All these factors make it the most economical choice.

The answer is (A).

72. When installing an irrigation system, the landscape architect will be required by local codes to use a backflow prevention device (local codes will specify the type), also known as a double-check device. A backflow preventer will keep any water within the system from entering a potable water supply, should a break and back-siphon situation occur.

The answer is (B).

73. The United States Department of Agriculture (USDA) soil classification system is used in regulating agricultural practices.

The American Association of State Highway and Transportation Officials (AASHTO) and the unified system deal with road and highway construction standards; they do not deal with horticultural practices related to growing and sustaining plantings. The American Society for Testing and Materials (ASTM) sets the standards for testing methods.

The answer is (C).

74. The proper sequencing of a construction project is as follows: (1) receiving bids and awarding the project, (2) performing mass earthwork, (3) installing minor site structures, and (4) installing irrigation and plants. Understanding construction sequencing is crucial to designing projects that can be bid, awarded, and constructed.

The answer is (B).

75. When evaluating the construction of a project, many factors need to be considered. In the given situation, reusing the existing asphalt section is the most economical and fastest method of preparing a road base. Distance from the nearest quarry, disposal costs, and aggregate costs all affect a project plan.

The answer is (B).

76. In a mass earthwork project, the most cost-effective and fastest method of moving materials is with a large box scraper. A box scraper can remove materials, store them, transport them to another location, and then re-spread them in small lifts, allowing for proper compaction of the materials as they are moved.

The answer is (B).

77. When staking a large-scale construction project, such as a state highway, a surveyor should use a global positioning system (GPS) unit. GPS units are portable, give accurate readings, and need not take line-of-sight measurements. Additionally, GPS staking can be done by a very small crew, or even by one person, and is therefore very economical.

The answer is (A).

78. When installing a drainage system in a golf course that needs to remain playable and free from site damage, a wheel trencher should be used. Wheel trenchers allow for quick trenching, clean operation, and quick system installation after trenching, thereby precluding the need for course closure while the work is being completed. Turf tires on the trencher will protect the greens and fairways from damage.

The answer is (C).

79. The least expensive fastener type to use in the construction process is the nail. Each type of fastener is intended for specific uses, which must be taken into account. But in terms of expense, nails are less costly than screws or bolts.

The answer is (A).

80. Northings and eastings are used to locate critical components on a construction project. For reasons related to record-keeping and safety, it is often necessary for the features of a construction project to be related to the surrounding infrastructure, utilities, buildings, roads, and distances.

The answer is (A).

81. A bearing is always taken with the north or south direction as the starting point, depending upon which hemisphere the surveyor is in. For example, most road and building projects use a northing and easting to locate critical construction points.

The answer is (C).

82. Preparation requirements for topsoil stripping and stockpiling would be found within the Division 31 specifications of the construction documents.

The general conditions of the construction documents cover information on project management by the general contractor. The contract closeout is the phase in which final pay applications are submitted, substantial completion documents are prepared, and final payments and retainage periods are begun. Planting notes would be found on the planting plan.

The answer is (B).

83. The primary role of the landscape architect during the construction observation phase of a project is to ensure compliance with the intent of the contract documents.

While it is also the landscape architect's job to address change orders and ensure specified means and methods are followed, it is not his or her responsibility to ensure that contract labor laws are followed.

The answer is (C).

84. Quality control is accomplished with on-site construction observation. The landscape architect does not administrate or direct a contractor during this phase. It is the landscape architect's role to observe, make notes, and return to the office or job shack to ensure proper compliance with the documents. If direction or additional information is required, that should always originate from the landscape architect's office, in an official document of record.

The answer is (B).

85. During the design and construction process, some items (such as irrigation plans) are left unspecified, or possibly shown graphically. The contractor is given latitude to use personal knowledge, experience, and methods of construction to achieve the end goal in these areas. Prior to final payment, the contractor is to submit a set of drawings showing all components and materials used for the item. These are known as as-built drawings. Shop drawings are submitted before fabrication, and as-built drawings are submitted after completion, as part of the final record drawings.

The answer is (D).

86. Survey control points are typically installed on a project prior to site clearing. These control points are critical to every aspect of a landscaping project, for both the contractor and the design team. They are essential for reasons of liability, for potential change orders, and for verification of all components of the project, from clearing, to grading, to building locations, to irrigation system connections.

The answer is (D).

Materials and Methods of Construction

87. A spike is a long, galvanized-metal nail that is driven down through a timber and into the timber or timbers below, to strengthen a structure and fasten the timber in place.

Wood screws large enough to be used in this application would be difficult to install. Wall ties are for use with masonry products, and a snap tie would be used for the formwork of a concrete wall or foundation.

The answer is (A).

88. Structural fill soils are used under floor slabs and building foundations, or in other situations where movement or settling of a subsoil must not occur. Structural fill is also used to bridge over poor soil conditions (that is, to transition or transfer weight or water) below any site components or elements.

The answer is (A).

89. Rebar is sized with a number that gives its diameter. To calculate the diameter, multiply the rebar's number by $1/8$ in [3.175 mm]. In the case of number 4 rebar, the diameter is $1/2$ in [1.25 cm].

The answer is (B).

90. In the United States, the standards for concrete are governed by the American Society for Testing and Materials (ASTM). The ASTM sets the standards and guidelines for many products used in the U.S. construction industry.

DIN and TUV are European standards and are based on the metric system; they are the European equivalents to the U.S. safety standards set forth by the ASTM. CSA is the Canadian Standards Association, the Canadian equivalent to the ASTM.

The answer is (C).

91. The geodetic survey method takes into account the curvature of the earth and is used for large-scale mapping and project planning. For example, the National Oceanic and Atmospheric Administration (NOAA) would use a geodetic survey to map a section of the Bering Sea.

The answer is (C).

92. When designing a deck, always consider the materials used and how long those materials will last. Also be aware that the framing design may be dictated by materials selection and the maintenance level that the owner can realistically sustain.

Staining is considered a maintenance procedure.

The answer is (C).

93. The information needed to properly size deck materials would be found in a wood span table. This type of table lists various materials available from different areas of the country.

ASTM standards govern the manufacture and testing of materials. Sieve analysis is used for soils, in the determination of footing size. Connector size tables would be used in this situation, but the information derived from them would be dependent on information obtained from wood span tables. Wood span tables will determine the deck material sizes needed, based upon the wood species and other aspects of the deck, such as the spans of joists and beams.

The answer is (C).

94. The design load for a deck would be done using pounds per square foot [kilograms per square meter]. Wood span tables use load calculations based upon pounds per square foot [kilograms per square meter].

Pounds [kilograms] is a weight measure. Feet [meters] is distance measure. Although pounds per square inch [kilograms per square centimeter] can be derived from pounds per square foot [kilograms per square meter], it is not the unit typically given in wood span tables.

The answer is (A).

95. The U.S. marketplace uses the footcandle to measure light. The lux is the international standard unit for lighting.

The answer is (C).

96. Irrigation system installation is typically one of the last sitework tasks to occur. It will be done after the completion of mass earthwork, asphalt paving, concrete work, building installation, and installation of site utilities. However, in order for the system to reach many locations on the site, irrigation sleeves must be installed so that pipes and wires can be added later without introducing the need to remove other items. This work must be done after the mass earthwork, but prior to installation of any permanent components. Close coordination of multiple contractors is required to ensure proper and timely installation.

The answer is (A).

97. When designing an irrigation system, the landscape architect should ensure that the same amount of water is distributed throughout a given zone. However, landscape sizes vary, and it would not be economical to break each zone out by size, radius, or the sprinkler type used, so manufacturers have created sprinkler heads that vary in radius and in flow rate. All these heads distribute water by matching precipitation rates (mpr).

The answer is (B).

98. Sizing the control wires is a key factor in irrigation design. To properly determine control wire size, the voltage drop must be calculated using Ohm's law. Ohm's law finds the voltage drop by multiplying the current (listed in the manufacturer's catalog for each valve) times the resistance for the type of wire used.

The answer is (A).

99. The strength of concrete comes from a direct relation of the cement and the aggregate found in the product. The amount of cement and the hardness of the aggregate combine to create the relative strength of that product.

Reinforcing will help the product carry a load, or not fail, but will not strengthen the mix. Too much water can actually weaken the mix.

The answer is (C).

100. The strength of concrete comes from a chemical process, and concrete gets harder as it ages. Concrete is considered to have reached full strength 28 days after mixing and placing.

The answer is (D).

101. When cement is mixed with water, a chemical process called hydration is started. Hydration is the growth of the cement particles with the addition of water. The particles continue to grow until they touch other particles or aggregates within the concrete mix. This process ultimately results in the strength, or hardness, of the design mix.

The answer is (B).

102. If extended time is required to transport, work, place, or finish concrete, a retardant should be added to the mix. A retardant is an admixture that slows down the hydration process in the concrete, allowing for additional time to work the product.

The answer is (D).

103. In landscaping projects, three main materials are used for masonry construction: brick, concrete, and stone. These materials can be manufactured or cut into various shapes, sizes, colors, and weights, and can form both structural and aesthetic items within a wide array of applications.

The answer is (D).

104. Masonry products offer excellent benefits to landscaping projects. Because masonry is durable, it lasts a long time and stands up to vandalism, and therefore can be safer than other products. And since it resists moisture, it will not rot. Its greater compressive strength and weight also contribute to longer-lasting structures with fewer maintenance needs.

The answer is (D).

105. With masonry block units, a solid, watertight seal is achieved through a compressed, tooled joint. The mason will install the mortar and then set the block into the mortar. Then, the mason will beat the block down into the mortar and clean the joint with a rounded tool, thereby compressing the remaining mortar into place, creating a solid, watertight seal.

The answer is (A).

106. Mortar is used in masonry construction to bond units together, but it also makes up for irregularities within individual units, allows for use of ties and reinforcing within the units, and combines all these functions with aesthetics, providing color and texture enhancement.

The answer is (D).

107. Wall ties are installed between masonry units to hold them in place. But wall ties also have tabs that extend perpendicularly to the masonry units and can be used for installing veneered clay brick units.

In creating a masonry unit with a brick veneer, steel rebar would be integral to the wall but would not hold the bricks. Anchors are for mounting wood or steel to a unit. Reinforcing wire is used between masonry units for strength.

The answer is (B).

108. Installing a railing into a sleeve is a common construction detail. This detail should be finished using a non-shrink grout, to secure the railing and prevent water from entering the unit.

The answer is (B).

109. Tensile strength is the ability of wood to resist stresses applied parallel to the wood grain. It is the greatest amount of longitudinal stress the lumber can tolerate prior to failure.

The answer is (C).

110. Lumber purchased in nominal sizes is first sized and then finished by drying and planing. Lumber will shrink when it is kiln dried.

The answer is (C).

111. Ledgers for decks are wood and are attached to the side of a new concrete wall or foundation with anchor bolts. The best installation method is to imbed the anchor bolt in the new concrete.

The answer is (D).

112. A compressive, or downward, force will reduce the size of construction materials such as soil or aggregates by compacting them. For example, 1 yd^2 of soil will go from 1 yd^2 as measured and sold from a supplier to $9/10$ yd^2 after compaction (a 10% compaction rate), resulting in a slightly lower finish grade than when it was first spread. This compaction rate must be realized when calculating volumes of soil on a project.

The answer is (C).

113. When the load on a wooden deck is exceeded, the deck's posts will be subjected to too much downward force, resulting in the crushing of the support posts.

An example of deflection would be if the joists were too far apart, causing the decking to sag.

The answer is (C).

114. The major load-carrying members in a deck are the beams.

The joists are either attached to the beams with joist hangers or set atop the beams, and their load is transferred to the beams. Decking, the surface material, transfers the load to the joists. Railings protect users from a fall.

The answer is (B).

115. A beam that spans more than two posts is considered a continuous-span beam.

If the beam hung out over a post and ended prior to connecting to another post, it would be a cantilever beam. All beams are loadbearing.

The answer is (B).

116. A geodetic survey takes into account the curvature of the earth. Geodetic surveys are critical to very large-scale projects in remote areas, such as the laying of oil lines or of roads that cross vast expanses of land.

The answer is (D).

117. Modern plane surveying uses a laser to verify and measure distances and elevations. The instrument stores the information digitally, downloading it into a program that is then converted into a topographic map including contours, elements located on site, property corners, and other pertinent information. Foresights and backsights are used to verify the surveyor's location, based on a known location.

The answer is (C).

118. Bearing capacity, a soil's ability to bear weight, is the critical soil characteristic that determines the size and shape of a footing design.

Soil classification systems establish the characteristics or structure of a soil, including density and content of materials. A geotechnical engineer can reference this data to calculate the bearing capacity.

The answer is (C).

119. Silt, sand, and clay are the three components that make up soil.

Bark is a wood product and, when decomposed, is a great additive to hold moisture in a soil. The amount of water held within a soil can be affected by adding decomposed wood products such as bark, sawdust, and compost.

The answer is (C).

120. The angle of repose is the stable slope formed when soil is placed naturally without compaction or manipulation. This angle may vary with soil type, but a 2:1 maximum slope is most common.

The answer is (B).

121. A soil is considered neutral when its pH rating is 6.6 to 7.3. Soil pH is critical to the survival of some plants and changes the color of others. In some cases, the selection of plant material should be based upon pH and drought conditions. Soil pH can also affect durability of elements such as rock and plastics used in landscape construction projects.

The answer is (C).

122. The plasticity of a soil is critical in determining where the soil may give way, causing the failure and possible collapse of a structure.

Cohesion is the ability of a soil to bind together under moderately dry conditions. Permeability is the ability of a soil to drain and remove water from the soil profile. Liquefaction is the point at which soil acts like a liquid, typically under earthquake conditions.

The answer is (C).

123. Soil plasticity is achieved when enough moisture leaves the soil profile and the soil is no longer in a liquid state. At this point the soil has some cohesive abilities and enters a plastic state.

The answer is (B).

124. A cantilever wall is a cast-in-place concrete retaining wall. This wall type is used for moderate to high walls and is very expensive to construct.

Dry-stack stone, gabions, crib walls, and bin walls are moderate-cost to low-cost retaining-wall types. Although these are very quick to build and inexpensive, they offer no choice of finish options to match particular design criteria.

The answer is (C).

125. "Sill" and "ledger" are construction terms related to framing; they are not forms of retaining walls.

There are three classifications of retaining walls: soil-bioengineering, gravity, and cantilevered. A soil-bioengineering wall is a reinforced embankment using geotextile materials, plantings, stone, or concrete surfacing. A gravity wall is a unit-stack wall system composed of materials such as stack stone or keystone block. Crib walls, bin walls, and gabions are all types of gravity walls. A cantilevered wall, also called a rigid wall, is solid concrete and cast-in-place. Cantilevered walls are the most costly retaining-wall type but have the greatest potential for height gain and offer options for the finish and design pattern.

The answer is (A).

126. It is acceptable to use weep holes with any rigid form of wall construction. A footing drain may also be used to relieve water pressure from behind a wall.

Precast masonry unit walls, dry-stack (rockery) walls, and gabion walls are not solid construction, so water can weep between the units or rocks. However, weep holes can also be installed in these wall types to help relieve pressure caused by excessive water behind the wall.

The answer is (A).

127. The lateral load of a retaining wall is the force of the soil being retained against the wall. Only soil that exceeds the angle of repose exerts pressure on a wall. This pressure can be calculated from the types of soil and the amounts of moisture the soils will retain.

The answer is (C).

128. The general contractor is required to keep the job site clean, safe, in proper order, and compliant with all local codes, laws, and safety standards. A well-maintained job site will also be in condition for proper inspection of work by municipal officials, the owner, and the design team.

The answer is (B).

129. Asphalt roadways always have a shoulder of aggregate, which helps improve the drainage of the asphalt surface. The extra gravel also supports the outside edge of the pavement, discouraging failure of the asphalt edge.

The answer is (C).

130. Vibration is used to consolidate concrete and fill all void spaces within a form. Voids in a foundation, wall, or footing will weaken the structure.

The answer is (B).

131. Concrete can be placed on a project in many ways, but the most common method is to cast it in place, pouring or pumping it directly on site. Walls, curbs, sidewalks, patios, footings, and so on are all forms of cast-in-place concrete.

Concrete can also arrive at the site dry, in bags, and be mixed on site. Finally, concrete can arrive on site in the form of precast units, such as block wall units, prestressed floors, or beams.

The answer is (A).

132. Cobblestone is not commonly used as an aggregate in a concrete mixture.

The three basic types of aggregates found in most standard concrete design mixes are sand, crushed stone, and gravel. Crushed stone can be made from gravel or from quarried rock.

The answer is (D).

133. Of the various methods of concrete testing, the slump test is the most common for on-site use. This test is used to measure the workability of the mix.

The answer is (C).

134. Prior to installing concrete on a project, the base must be properly prepared and tested for compaction, to avoid settling of the concrete. The subgrade is tested using the Proctor test method, as described by the American Society for Testing and Materials (ASTM).

The answer is (B).

135. Surveying is critical for proper construction on a landscaping project. Plane surveying is the method typically used on landscape architecture projects. Unlike a geodetic survey, which uses a coordinate system for locating points on the earth, a plane survey disregards the earth's curvature and is considered an ordinary field-survey method for projects of small to moderate size.

Metes and bounds is a type of legal description of a site or lot.

The answer is (B).

136. Mortar is composed of sand, cement, and water. If gravel were added, the mixture would be called concrete.

The answer is (C).

137. Unseasoned, or "green," lumber is wood that has been freshly cut and has not been cured, either by a kiln or by air drying under cover.

The answer is (A).

138. The moisture content of lumber is marked on the lumber itself, adjacent to the "kiln dried" stamp. This mark is important to ensure that wood used for finish carpentry or indoor applications will not continue to warp, shrink, or change in shape and size.

The answer is (C).

139. Lumber is purchased in lumber yards by specific lengths and sizes and comes from the mill precut to these industry-standard units. However, lumber as a commodity is sold by the board foot and then milled into various sizes.

The answer is (C).

140. Various parcels can be combined, and the resulting property lines can be adjusted by means of a boundary line adjustment, or consolidation, which must be performed by a professional land surveyor. A boundary line adjustment can improve the design of a project, as well as other aspects such as economics, by adjusting the design to satisfy local codes and minimize safety impacts.

The answer is (D).

141. Construction adhesive offers several benefits for a deck construction project: it will bond the wood together; it can be used on dry, wet, or treated wood; and it is not affected by moisture. When applied in a consistent bead it will help create a moisture barrier.

Not all wood glues can be used in an exterior environment. Mortar would be used in masonry construction. A bituminous binder is used in asphalt construction only, to bond old and new asphalt.

The answer is (C).

142. The compaction of soils is a critical part of the construction process. There are many types of compactors for various jobs and soil conditions. For a colloidal soil condition, the best compactor choice is a sheep's foot roller. Sheep's foot rollers are used exclusively with cohesive soils, to allow deep compaction of the materials.

Plate compactors are best suited for aggregate products. Walk-behind tampers are for small trench work, and a double-drum vibrating roller is appropriate for asphalt work.

The answer is (C).

143. A tack coat is used in asphalt paving to adhere two lifts of asphalt together.

Primer is a product used for the initial overlayment of the first course of asphalt. A binder, which is a type of tack coating, is used to adhere an old and new coat of asphalt. Expoxies are used in the construction of wood or concrete to anchor or attach products.

The answer is (B).

144. Welded wire mesh is used to reinforce concrete flatwork, for increased slab tensile strength.

A construction joint is a break that occurs between concrete pours or sections during installation. A contraction joint is installed to control cracking of the concrete due to shrinkage and swelling, which are caused by exposure to heat or cold. Additives allow the concrete to be worked and installed under various conditions, such as hot or cold weather, or under water.

The answer is (B).

145. The depth of a contraction joint should be no less than 1/6 the depth of the concrete slab.

1/2 in [12.7 mm] would not be deep enough placement to control cracking in a slab. A depth that matched the maximum size of the aggregate would weaken the slab by removing 1/3 of the concrete thickness, since surface mortar is generally removed to a depth of 1/3 the diameter of the aggregate particles.

The answer is (C).

146. The hardness of the rock will affect the wear of the rock in concrete.

Availability, aggregate size, and aggregate quantity in the mix are all aspects of concrete mix components that affect the cost of the product but do not affect the wearing ability of the aggregate.

The answer is (A).

147. Certain concrete mixes, used for specific needs or purposes, call for the addition of materials other than cement, aggregate, or water. These materials are called admixtures and are used to control hardening and workability, and to boost the cementing properties of the mix.

The answer is (C).

148. In poor soil conditions or areas where liquefaction can occur, only low-ground-pressure, tracked construction equipment is appropriate.

With wheeled equipment, all the weight is borne in a very small area, which creates ruts in the traveled landscape. A skid-steer loader functions by counter-rotation; the left and right sides spin in opposite directions, which turns and tears up soft, wet ground. Drag-line buckets are used in underwater situations.

The answer is (D).

149. When using a common nail to install decking materials, choose a nail that is roughly 2 1/2 times the thickness of the wood member being secured. The nail must penetrate the depth of the wood member and then drive into the framing materials for a secure hold.

The answer is (C).

150. *Customary U.S. Solution*

Footing area takes into account the bearing capacity of the soil.

$$A_{\text{min}} = \frac{\text{deck load}}{\text{soil bearing capacity}} = \frac{4000 \text{ lbf}}{7 \frac{\text{lbf}}{\text{in}^2}} = 571.43 \text{ in}^2$$

This is equivalent to a square with sides of

$$s = \sqrt{A_{\text{min}}} = \frac{\sqrt{571.43 \text{ in}^2}}{12 \frac{\text{in}}{\text{ft}}} = 1.99 \text{ ft}$$

So a 2 ft × 2 ft footing meets the minimum requirement.

SI Solution

Footing area takes into account the bearing capacity of the soil.

$$A_{\text{min}} = \frac{\text{deck load}}{\text{soil bearing capacity}} = \frac{1800 \text{ kg}}{5000 \frac{\text{kg}}{\text{m}^2}} = 0.36 \text{ m}^2$$

This is equivalent to a square with sides of

$$s = \sqrt{A_{\text{min}}} = \sqrt{0.36 \text{ m}^2} = 0.6 \text{ m}$$

So a 0.6 m × 0.6 m footing meets the minimum requirement.

The answer is (C).

151. Decking is typically installed bark side up. This orientation allows for cupping, or warping, of the decking materials as they shrink and swell over time, without creating a tripping hazard. Decking materials are not all laid flat; that is a design issue.

The answer is (C).

152. Installing a proper ground wire—one that has been welded to the grounding rod or plate and then connected to the lighting control panel—will help ensure system safety in the case of a lighting strike or power surge.

Connecting the panel to a downspout or to metal conduit with incoming wires will not adequately disperse the energy from a lightning strike or power surge. Proper grounding is a requirement.

The answer is (B).

153. Graded gravels are the best structural fill materials. They have a bridging factor that allows them to spread out over poor bearing soils.

Silts are moisture sensitive and are difficult to compact except under pristine conditions. Sands are considered to be noncohesive soils. Crushed aggregates can be used as structural fill materials, but they are typically not cost effective in this application.

The answer is (C).

154. DBH stands for diameter breast height and is the measure used to specify trees and determine a proper size per the specifications. Trees that fail to meet the DBH industry standards for nursery stock should be rejected from the project and not installed.

The answer is (B).

155. A gabion wall is constructed of crushed rock and wire baskets. Gabion walls are easy to construct and convenient for non-aesthetic solutions where cost is a major factor of the project.

The answer is (C).

156. Batter is the slight tilting of a retaining wall into the slope that the wall is designed to retain. Both the exposed face and back of a wall can have a batter, depending upon the structural design.

The answer is (B).

157. Geotextile fabric is commonly misapplied. Furthermore, over time the fabric can become clogged or obstructed with fine particulate matter, which creates a thin, impervious soil lens. This soil lens can decrease, or even eliminate, the infiltration function of the rain garden. The diminished infiltration function of the soil profile would result in the failure of the rain garden as a stormwater mitigation system, and is therefore inappropriate for installation.

The answer is (D).

References

Harris, Charles W., and Nicholas T. Dines. *Time-Saver Standards for Landscape Architecture*. McGraw-Hill.

Lane, Kenneth F., and John Roberts. *Fundamental Land: A Workbook of Topography, Grading, Drainage, Earth Volumes, Areas, Retaining Walls, and Road and Walk Design*. Iowa State University.

Nelischer, Maruice, ed. *Handbook of Landscape Architectural Construction*. American Society of Landscape Architects.

Ramsey, Charles G., et al. *Architectural Graphic Standards*. John Wiley and Sons.

Strom, Steven, and Kurt Nathan. *Site Engineering for Landscape Architects*. John Wiley and Sons.